tredition®

tredition was established in 2006 by Sandra Latusseck and Soenke Schulz. Based in Hamburg, Germany, tredition offers publishing solutions to authors and publishing houses, combined with worldwide distribution of printed and digital book content. tredition is uniquely positioned to enable authors and publishing houses to create books on their own terms and without conventional manufacturing risks.

For more information please visit: www.tredition.com

TREDITION CLASSICS

This book is part of the TREDITION CLASSICS series. The creators of this series are united by passion for literature and driven by the intention of making all public domain books available in printed format again - worldwide. Most TREDITION CLASSICS titles have been out of print and off the bookstore shelves for decades. At tredition we believe that a great book never goes out of style and that its value is eternal. Several mostly non-profit literature projects provide content to tredition. To support their good work, tredition donates a portion of the proceeds from each sold copy. As a reader of a TREDITION CLASSICS book, you support our mission to save many of the amazing works of world literature from oblivion. See all available books at www.tredition.com.

 Project Gutenberg

The content for this book has been graciously provided by Project Gutenberg. Project Gutenberg is a non-profit organization founded by Michael Hart in 1971 at the University of Illinois. The mission of Project Gutenberg is simple: To encourage the creation and distribution of eBooks. Project Gutenberg is the first and largest collection of public domain eBooks.

The Splash of a Drop

A. M. (Arthur Mason) Worthington

Imprint

This book is part of TREDITION CLASSICS

Author: A. M. (Arthur Mason) Worthington
Cover design: Buchgut, Berlin – Germany

Publisher: tredition GmbH, Hamburg - Germany
ISBN: 978-3-8472-1353-6

www.tredition.com
www.tredition.de

Copyright:
The content of this book is sourced from the public domain.

The intention of the TREDITION CLASSICS series is to make world literature in the public domain available in printed format. Literary enthusiasts and organizations, such as Project Gutenberg, worldwide have scanned and digitally edited the original texts. tredition has subsequently formatted and redesigned the content into a modern reading layout. Therefore, we cannot guarantee the exact reproduction of the original format of a particular historic edition. Please also note that no modifications have been made to the spelling, therefore it may differ from the orthography used today.

THE ROMANCE OF SCIENCE

THE SPLASH OF A DROP

BY
Prof. A.M. WORTHINGTON, M.A., F.R.S.

Being the reprint of a Discourse delivered at the Royal Institution of Great Britain, May 18, 1894.

PUBLISHED UNDER THE DIRECTION OF THE GENERAL
LITERATURE COMMITTEE.

LONDON:
SOCIETY FOR PROMOTING CHRISTIAN KNOWLEDGE,
NORTHUMBERLAND AVENUE, CHARING CROSS, W.C.;
43, QUEEN VICTORIA STREET, E.C.
Brighton: 129, NORTH STREET.
New York: E. & J.B. YOUNG & CO.
1895.

THE SPLASH OF A DROP

INSTANTANEOUS PHOTOGRAPHS OF THE SPLASH OF A WATER-DROP FALLING ABOUT 16 INCHES INTO MILK.

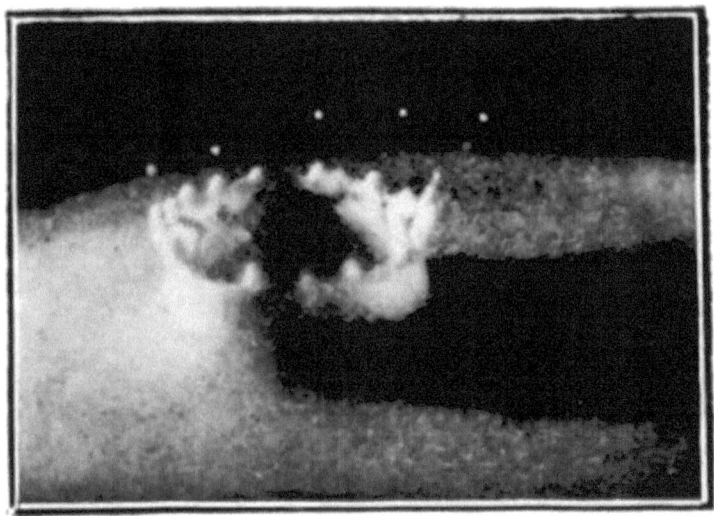

Time after contact = .0262 sec.

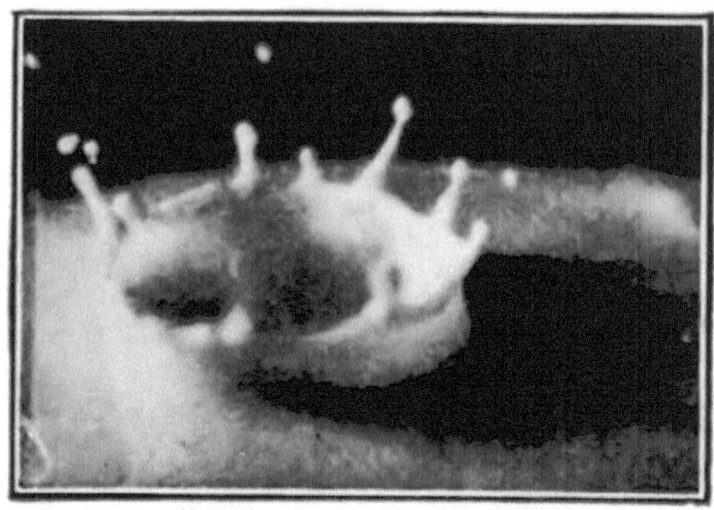

Time after contact = ·0391 sec.

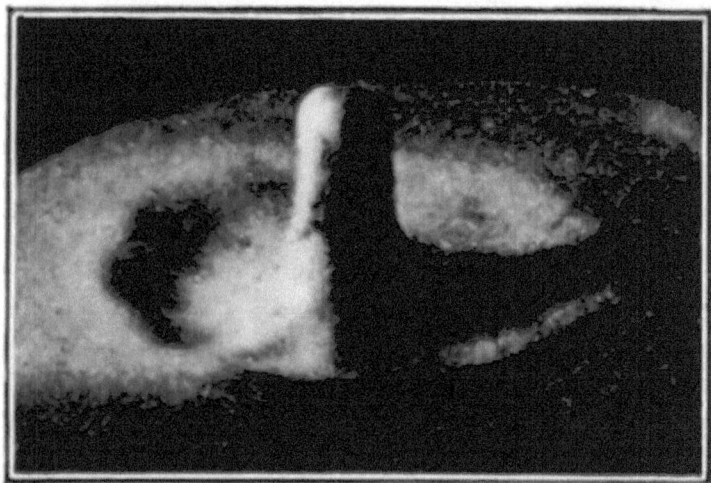

Time after contact = ·101 sec.

THE SPLASH OF A DROP

The splash of a drop is a transaction which is accomplished in the twinkling of an eye, and it may seem to some that a man who proposes to discourse on the matter for an hour must have lost all sense of proportion. If that opinion exists, I hope this evening to be able to remove it, and to convince you that we have to deal with an exquisitely regulated phenomenon, and one which very happily illustrates some of the fundamental properties of fluids. It may be mentioned also that the recent researches of Lenard in Germany and J.J. Thomson at Cambridge, on the curious development of electrical charges that accompanies certain kinds of splashes, have invested with a new interest any examination of the mechanics of the phenomenon. It is to the mechanical and not to the electrical side of the question that I shall call your attention this evening.

The first well-directed and deliberate observations on the subject that I am acquainted with were made by a school-boy at Rugby some twenty years ago, and were reported by him to the Rugby Natural History Society. He had observed that the marks of accidental splashes of ink-drops that had fallen on some smoked glasses with which he was experimenting, presented an appearance not easy to account for. Drops of the same size falling from the same height had made always the same kind of mark, which, when carefully examined with a lens, showed that the smoke had been swept away in a system of minute concentric rings and fine striæ. Specimens of such patterns, obtained by letting drops of mercury, alcohol, and water fall on to smoked glass, are thrown on the screen, and the main characteristics are easily recognized. Such a pattern corresponds to the footprints of the dance that has been performed on the surface, and though the drop may be lying unbroken on the plate, it has evidently been taking violent exercise, and were our vision acute enough we might observe that it was still palpitating after its exertions.

A careful examination of a large number of such footprints showed that any opinion that could be formed therefrom of the nature of the motion of the drop must be largely conjectural, and it occurred to me about eighteen years ago to endeavour by means of the illumination of a suitably-timed electric spark to watch a drop through its various changes on impact.

The reason that with ordinary continuous light nothing can be satisfactorily seen of the splash, is not that the phenomenon is of such short duration, but because the changes are so rapid that before the image of one stage has faded from the eye the image of a later and quite different stage is superposed upon it. Thus the resulting impression is a confused assemblage of all the stages, as in the photograph of a person who has not sat still while the camera was looking at him. The problem to be solved experimentally was therefore this: to let a drop of definite size fall from a definite height in comparative darkness on to a surface, and to illuminate it by a flash of exceedingly short duration at any desired stage, so as to exclude all the stages previous and subsequent to the one thus picked out. The flash must be bright enough for the image of what is seen to remain long enough on the eye for the observer to be able to attend to it, and even to shift his attention from one part to another, and thus to make a drawing of what is seen. If necessary the experiment must be capable of repetition, with an exactly similar drop falling from exactly the same height, and illuminated at exactly the same stage. Then, when this stage has been sufficiently studied, we must be able to arrange with another similar drop to illuminate it at a rather later stage, say 1/1000 second later, and in this way to follow step by step the whole course of the phenomenon.

The apparatus by which this has been accomplished is on the table before you. Time will not suffice to explain how it grew out of earlier arrangements very different in appearance, but its action is very simple and easy to follow by reference to the diagram (Fig. 1).

AA´ is a light wooden rod rather longer and thicker than an ordinary lead pencil, and pivoted on a horizontal axle O. The rod bears at the end A a small deep watch-glass, or segment of a watch-glass, whose surface has been smoked, so that a drop even of water will lie on it without adhesion. The end A´ carries a small strip of tinned iron, which can be pressed against and held down by an electro-magnet CC´. When the current of the electro-magnet is cut off the iron is released, and the end A´ of the rod is tossed up by the action of a piece of india-rubber stretched catapult-wise across two pegs at

E, and by this means the drop resting on the watch-glass is left in mid-air free to fall from rest.

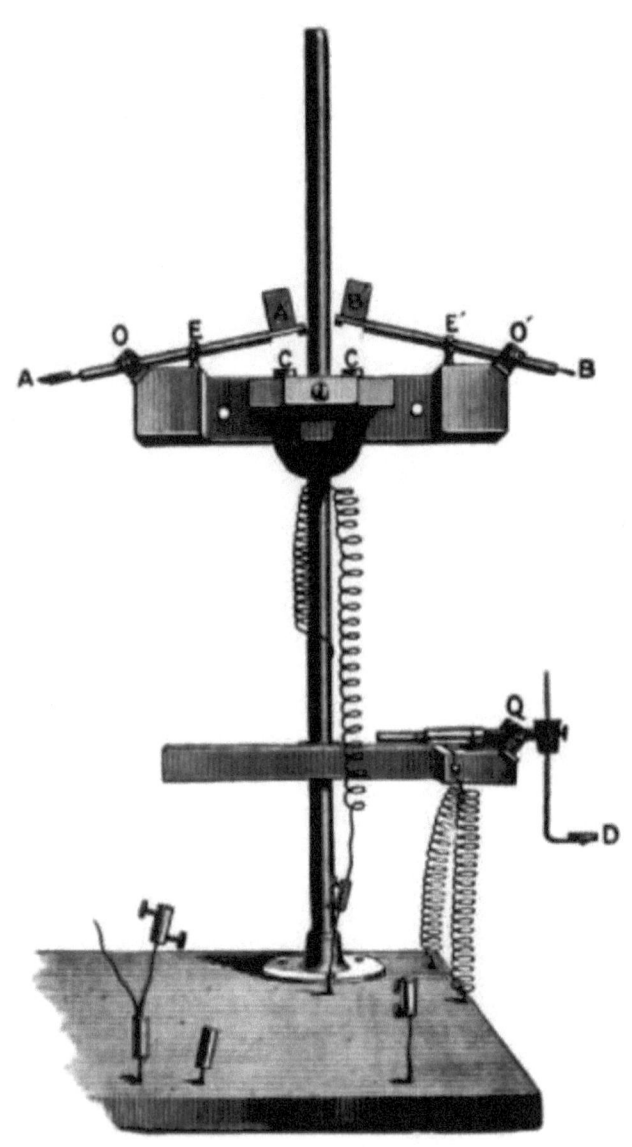

Fig. 1

BB′ is a precisely similar rod worked in just the same way, but carrying at B a small horizontal metal ring, on which an ivory timing sphere of the size of a child's marble can be supported. On cutting off the current of the electro-magnet the ends A′ and B′ of the two levers are simultaneously tossed up by the catapults, and thus drop and sphere begin to fall at the same moment. Before, however, the drop reaches the surface on which it is to impinge, the timing sphere strikes a plate D attached to one end of a third lever pivoted at Q, and thus breaks the contact between a platinum wire bound to the underside of this lever and another wire crossing the first at right angles. This action breaks an electric current which has traversed a second electro-magnet F (Fig. 2), and releases the iron armature N of the lever NP, pivoted at P, thus enabling a strong spiral spring G to lift a stout brass wire L out of mercury, and to break at the surface of the mercury a strong current that has circulated round the primary circuit of a Ruhmkorff's induction coil; this produces at the surface of the mercury a bright self-induction spark in the neighbourhood of the splash, and it is by this flash that the splash is viewed. The illumination is greatly helped by surrounding the place where the splash and flash are produced by a white cardboard enclosure, seen in Fig. 2, from whose walls the light is diffused.

Fig. 2.

It will be observed that the time at which the spark is made will depend upon the distance that the sphere has to fall before striking the plate D, for the subsequent action of demagnetizing F and pulling the wire L out of the mercury in the cup H is the same on each occasion. The modus operandi is consequently as follows: — The observer, sitting in comparative but by no means complete darkness, faces the apparatus as it appears in Fig. 2, presses down the ends A´B´ of the levers first described, so that they are held by the electro-magnet C (Fig. 1); then he presses the lever NP down on the electro-magnet F, sets the timing sphere and drop in place, and then by means of a bridge between two mercury cups, short-circuits and thus cuts off the current of the electro-magnet C. This lets off drop and sphere, and produces the flash. The stage of the phenomenon that is thus revealed having been sufficiently studied by repetition of the experiment as often as may be necessary, he lowers the plate D a fraction of an inch and thus obtains a later stage. Not only is any desired stage of the phenomenon thus easily brought under examination, but the apparatus also affords the means of measuring the time interval between any two stages. All that is necessary is to know the distance that the timing sphere falls in the two cases. Elementary dynamics then give us the interval required. Thus, if the sphere falls one foot and we then lower D 1/4 inch, the interval between the corresponding stages will be about ·0026 second.

Having thus described the apparatus, which I hope shortly to show you in action, I pass to the information that has been obtained by it.

This is contained in a long series of drawings, of which a selection will be presented on the screen. The First Series that I have to show represents the splash of a drop of mercury 0·15 inch in diameter that has fallen 3 inches on to a smooth glass plate. It will be noticed that very soon after the first moment of impact, minute rays are shot out in all directions on the surface. These are afterwards overflowed or united, until, as in Fig. 8, the outline is only slightly rippled. Then (Fig. 9) main rays shoot out, from the ends of which in some cases minute droplets of liquid would split off, to be left lying in a circle on the plate, and visible in all subsequent stages. By counting these droplets when they were thus left, the number of rays was ascertained to have been generally about 24. This exquisite shell-like configuration, shown in Fig. 9, marks about the maximum spread of the liquid, which, subsiding in the middle, afterwards flows into an annulus or rim with a very thin central film, so thin, in fact, as often to tear more or less irregularly. This annular rim then divides or segments (Figs. 14, 15, 16) in such a manner as to join up the rays in pairs, and thus passes into the 12-lobed annulus of Fig. 16. Then the whole contracts, but contracts most rapidly between the lobes, the liquid then being driven into and feeding the arms, which follow more slowly. In Fig. 21 the end of this stage is reached, and now the arms continuing to come in, the liquid rises in the centre; this is, in fact, the beginning of the rebound of the drop from the plate. In the case before us the drops at the ends of the arms now break off (Fig. 25), while the central mass rises in a column which just fails itself to break up into drops, and falls back into the middle of the circle of satellites which, it will be understood, may in some cases again be surrounded by a second circle of the still smaller and more numerous droplets that split off the ends of the rays in Fig. 9. The whole of the 30 stages described are accomplished in about 1/20 second, so that the average interval between them is about 1/600 second.

FIRST SERIES.

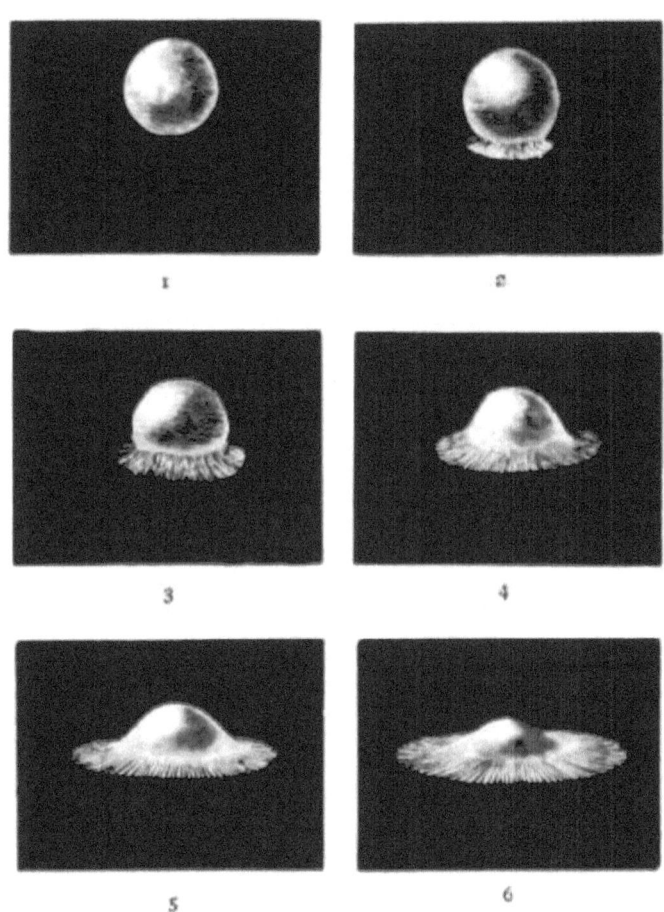

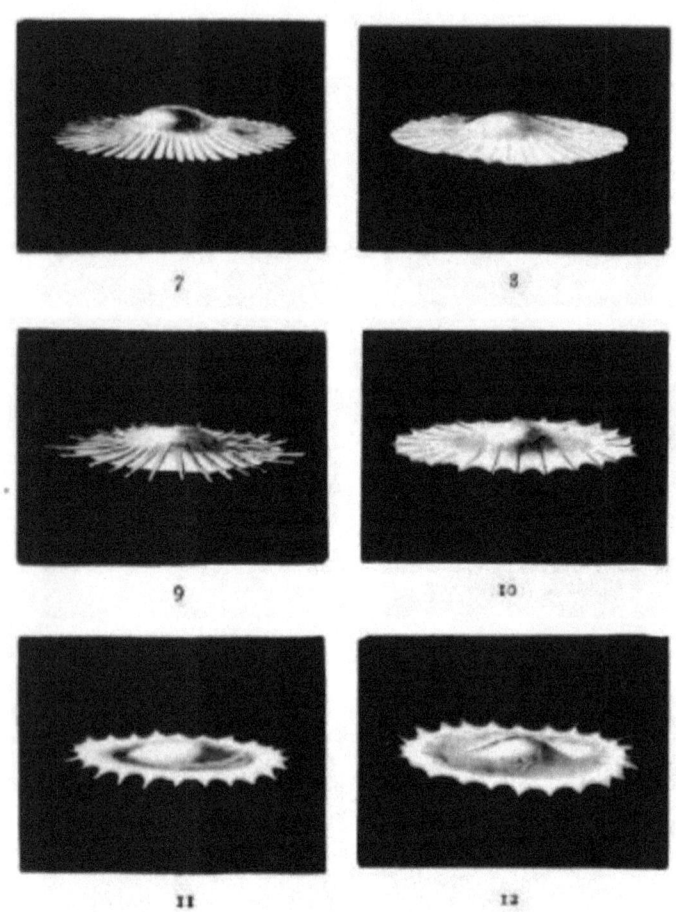

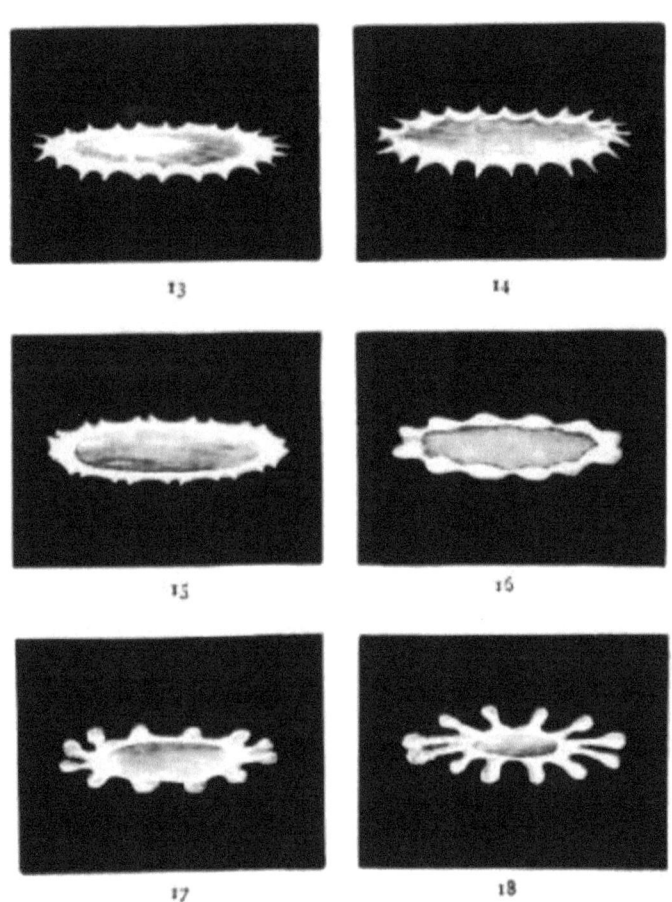

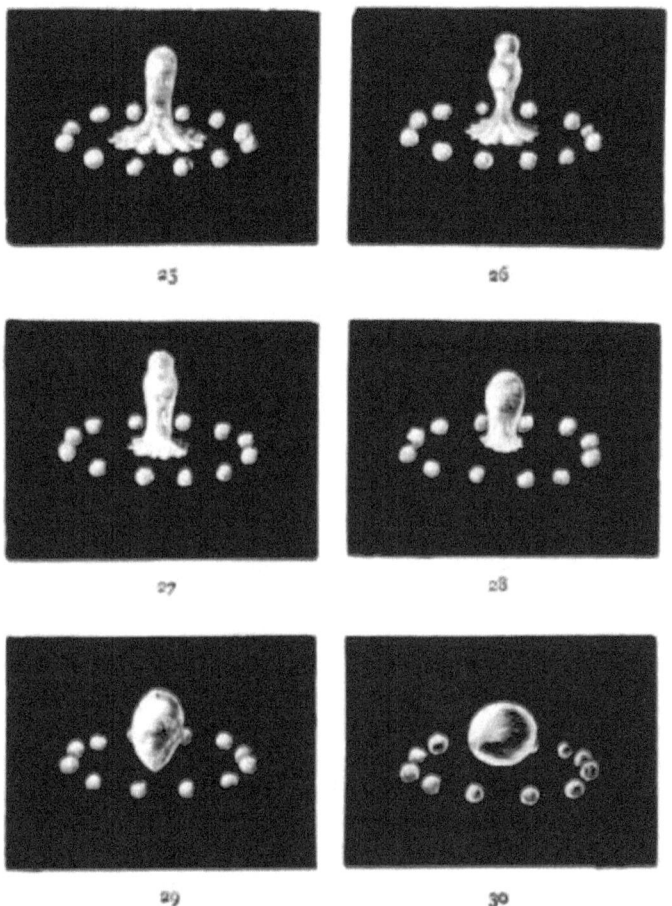

25 26

27 28

29 30

It should be mentioned that it is only in rare cases that the subordinate drops seen in the last six figures, are found lying in a very complete circle after all is over, for there is generally some slight disturbing lateral velocity which causes many to mingle again with the central drop, or with each other. But even if only half or a quarter of the circle is left, it is easy to estimate how many drops, and therefore how many arms there have been. It may be mentioned that sometimes the surface of the central lake of liquid (Figs. 14, 15, 16, 17) was seen to be covered with beautiful concentric ripples, not shown in the figures.

The question now naturally presents itself, Why should the drop behave in this manner? In seeking the answer it will be useful to ask ourselves another question. What should we have expected the drop to do? Well, to this I suppose most people would be inclined, arguing from analogy with a solid, to reply that it would be reasonable to expect the drop to flatten itself, and even very considerably flatten itself, and then, collecting itself together again, to rebound, perhaps as a column such as we have seen, but not to form this regular system of rays and arms and subordinate drops.

Now this argument from analogy with a solid is rather misleading, for the forces that operate in the case of a solid sphere that flattens itself and rebounds, are due to the bodily elasticity which enables it not only to resist, but also to recover from any distortion of shape or shearing of its internal parts past each other. But a liquid has no power of recovering from such internal shear, and the only force that checks the spread, and ultimately causes the recovery of shape, is the *surface tension*, which arises from the fact that the surface layers are always in a state of extension and always endeavouring to contract. Thus we are at liberty when dealing with the motions of the drop to think of the interior liquid as not coherent, provided we furnish it with a suitable elastic skin. Where the surface skin is sharply curved outwards, as it is at the sharp edge of the flattened disc, there the interior liquid will be strongly pressed back. In fact the process of flattening and recoil is one in which energy of motion is first expended in creating fresh liquid surface, and subsequently recovered as the surface contracts. The transformation is, however, at all moments accompanied by a great loss of energy as heat. Moreover, it must be remembered that the energy expended in creating the surface of the satellite drops is not restored if these remain permanently separate. Thus the surface tension explains the recoil, and it is also closely connected with the formation of the subordinate rays and arms. To explain this it is only necessary to remind you that a liquid cylinder is an unstable configuration. As you know, any fine jet becomes beaded and breaks into drops, but it is not necessary that there should be any flow of liquid along the jet;

if, for example, we could realize a rod of liquid of the shape and size of this cylindrical ruler that I hold in my hand, and liberate it in the air, it would not retain its cylindrical shape, but would segment or divide itself up into a row of drops regularly disposed according to a definite and very simple numerical law, viz. that the distances between the centres of contiguous drops would be equal to the circumference of the cylinder. This can be shown by calculation to be a consequence of the surface tension, and the calculation has been closely verified by experiment. If the liquid cylinder were liberated on a plate, it would still topple into a regular row of drops, but they would be further apart; this was shown by Plateau. Now imagine the cylinder bent into an annulus. It will still follow the same law, [1] *i.e.* it will topple into drops just as if it were straight. This I can show you by a direct experiment. I have here a small thick disc of iron, with an accurately planed face and a handle at the back. In the face is cut a circular groove, whose cross section is a semi-circle. I now lay this disc face downwards on the horizontal face of the lantern condenser, and through one of two small holes bored through to the back of the disc I fill the groove with quicksilver. Now, suddenly lifting the disc from the plate, I release an annulus of liquid, which splits into the circle of very equal drops which you see projected on the screen. You will notice that the main drops have between them still smaller ones, which have come from the splitting up of the thin cylindrical necks of liquid which connected the larger drops at the last moment.

Now this tendency to segment or topple into drops, whether of a straight cylinder or of an annulus, is the key to the formation of the arms and satellites, and indeed to much that happens in all the splashes that we shall examine. Thus in Fig. 12 we have an annular rim, which in Figs. 13 and 14 is seen to topple into lobes by which the rays are united in pairs, and even the special rays that are seen in Fig. 9 owe their origin to the segmentation of the rim of the thin disc into which the liquid has spread. The proceeding is probably exactly analogous to what takes place in a sea wave that curls over in calm weather on a slightly sloping shore. Any one may notice how, as it curls over, the wave presents a long smooth edge, from which at a given instant a multitude of jets suddenly shoot out, and at once the back of the wave, hitherto smooth, is seen to be furrowed or "combed." There can be no doubt that the cylindrical edge topples into alternate convexities and concavities; at the former the flow is helped, at the latter hindered, and thus the jets begin, and special lines of flow are determined. In precisely the same way the previously smooth circular edge of Fig. 8 topples, and determines the rays and lines of flow of Fig. 9.

Before going on to other splashes I will now endeavour to reproduce a mercury splash of the kind I have described, in a manner that shall be visible to all. For this purpose I have reduplicated the apparatus which you have seen, and have it here so arranged that I can let the drop fall on to the horizontal condenser plate of the lantern, through which the light passes upwards, to be afterwards thrown upon this screen. The illuminating flash will be made inside the lantern, where the arc light would ordinarily be placed. I have now set a drop of mercury in readiness and put the timing sphere in place, and now if you will look intently at the middle of the screen I will darken the room and let off the splash. (The experiment was repeated four or five times, and the figures seen were like those of Series X.) Of course all that can be shown in this way is the outline, or rather a horizontal section of the splash; but you are able to recognize some of the configurations already described, and will be the more willing to believe that a momentary view is after all sufficient to give much information if one is on the alert and has acquired skill by practice.

The general features of the splash that we have examined are not merely characteristic of the liquid mercury, but belong to all splashes of a liquid falling on to a surface which it does not wet, provided the height of fall or size of the drop are not so great as to cause complete disruption, [2] in which case there is no recovery and rebound. Thus a drop of milk falling on to smoked glass will, if the height of fall and size of drop are properly adjusted, give forms very similar to those presented by a drop of mercury. The whole course of the phenomenon depends, in fact, mainly on four quantities only: (1) the size of the drop; (2) the height of fall; (3) the value of the surface tension; (4) the viscosity of the liquid.

The next series of drawings illustrates the splash of a drop of water falling into water.

In order the better to distinguish the liquid of the original drop from that into which it falls, the latter was coloured with ink or with an aniline dye, and the drop itself was of water rendered turbid with finely-divided matter in suspension. Finally drops of milk were found to be very suitable for the purpose, the substitution of milk for water not producing any observable change in the phenomenon.

In Series II. the drop fell 3 inches, and was 1/5 inch in diameter.

[In most of the figures of this and of succeeding series the central white patch represents the original drop, and the white parts round it represent those raised portions of the liquid which catch the light. The numbers under each figure give the time interval in seconds from the occurrence of the first figure, or of the figure marked $\tau = 0$.]

SERIES II.

The Splash of a Drop, followed in detail by Instantaneous Illumination.

Diameter of Drop, 1/5 inch. Height of Fall, 3-1/5 inches.

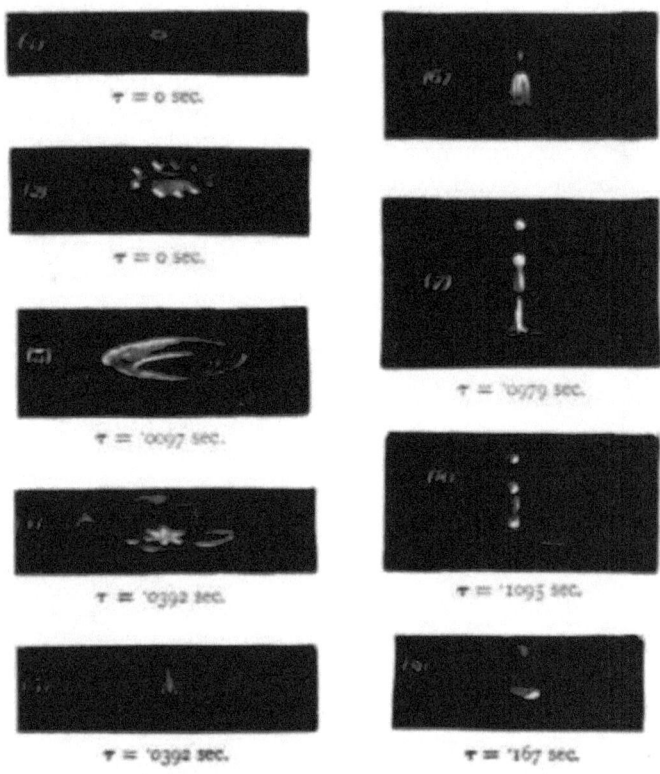

It will be observed that the drop flattens itself out somewhat, and descends at the bottom of a hollow with a raised beaded edge (Fig. 2). This edge would be smooth and circular but for the instability which causes it to topple into drops. As the drop descends the hollow becomes wider and deeper, and finally closes over the drop (Fig. 3), which, however, soon again emerges as the hollow flattens out, appearing first near, but still below the surface (Fig. 4), in a flattened, lobed form, afterwards rising as a column somewhat mixed with adherent water, in which traces of the lobes are at first very visible.

The rising column, which is nearly cylindrical, breaks up into drops before or during its subsequent descent into the liquid. As it disappears below the surface the outward and downward flow causes a hollow to be again formed, up the sides of which an annulus of milk is carried, while the remainder descends to be torn again a second time into a vortex ring, which, however, is liable to disturbance from the falling in of the drops which once formed the upper part of the rebounding column.

It is not difficult to recognize some features of this splash without any apparatus beyond a cup of tea and a spoonful of milk. Any drinker of afternoon tea, after the tea is poured out and before the milk is put in, may let the milk fall into it drop by drop from one or two inches above it. The rebounding column will be seen to consist almost entirely of milk, and to break up into drops in the manner described, while the vortex ring, whose core is of milk, may be seen to shoot down into the liquid. But this is better observed by dropping ink into a tumbler of clear water.

Let us now increase the height of fall to 17 inches. Series III. exhibits the result. All the characteristics of the last splash are more strongly marked. In Fig. 1 we have caught sight of the little raised rim of the hollow before it was headed, but in Fig. 2 special channels of easiest flow have been already determined. The number of ribs and rays in this basket-shaped hollow seemed to vary a good deal with different drops, as also did the number of arms and lobes seen in later figures, in a somewhat puzzling manner, and I made no attempt to select drawings which are in agreement in this respect. It will be understood that these rays contain little or none of the liquid of the drop, which remains collected together in the middle. Drops from these rays or from the larger arms and lobes of subsequent figures are often thrown off high into the air. In Figs. 3 and 4 the drop is clean gone below the surface of the hollow, which is now deeper and larger than before. The beautiful beaded annular edge then subsides, and in Fig. 5 we see the drop again, and in Fig.

6 it begins to emerge. But although the drop has fallen from a greater height than in the previous splash, the energy of the impact, instead of being expended in raising the same amount of liquid to a greater height, is now spent in lifting a much thicker adherent column to about the same height as in the last splash. There was sometimes noticed, as seen in Fig. 9, a tendency in the water to flow up past the milk, which, still comparatively unmixed with water, rides triumphant on the top of the emergent column. The greater relative thickness of this column prevents it splitting into drops, and Figs. 10 and 11 show it descending below the surface to form the hollow of Fig. 12, up the sides of which an annular film of milk is carried (Figs. 12 and 13), having been detached from the central mass, which descends to be torn again, this time centrally into a well-marked vortex ring.

SERIES III.

The Splash of a Drop, followed in detail by Instantaneous Illumination.

Diameter of Drop, 1/5 inch. Height of Fall, 1 ft. 5 in.

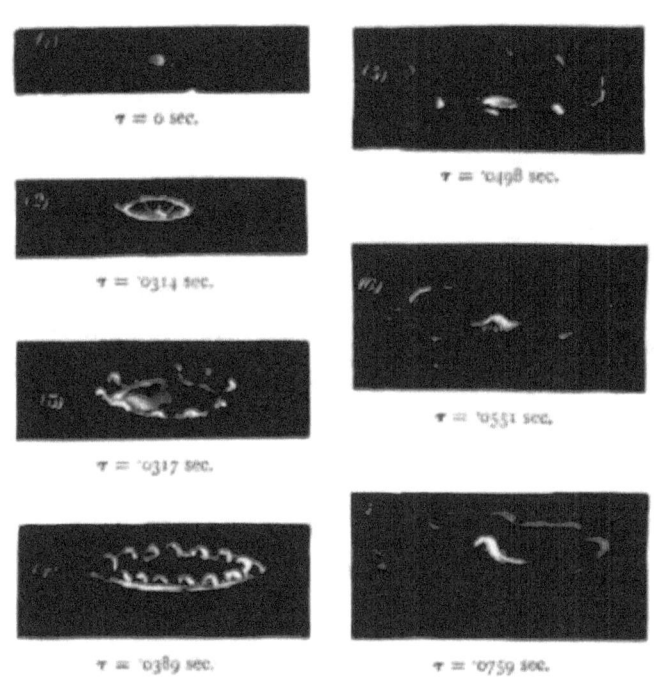

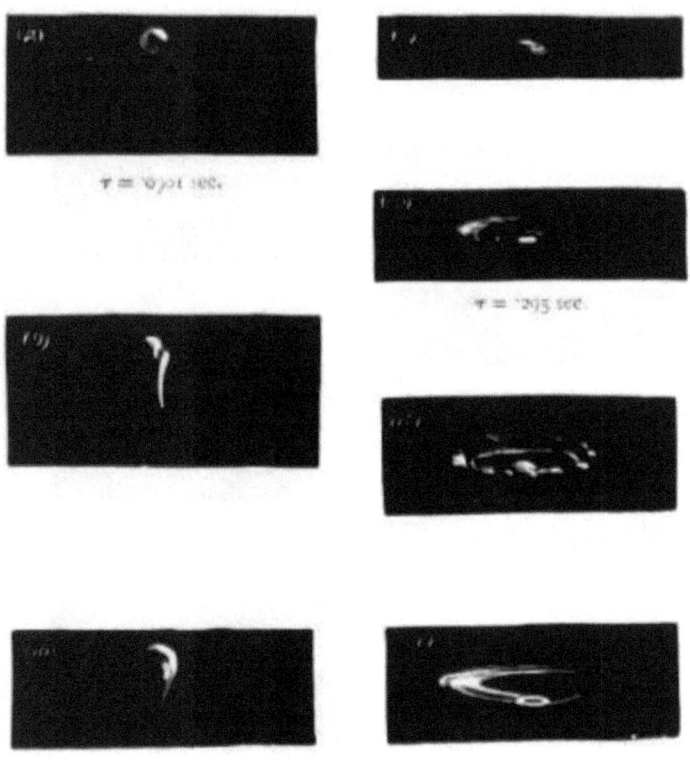

If we keep to the same size of drop and increase the fall to something over a yard, no great change occurs in the nature of the splash, but the emergent column is rather higher and thinner and shows a tendency to split into drops.

When, however, we double the volume of the drop and raise the height of fall to 52 inches, the splash of Series IV. is obtained, which is beginning to assume quite a different character. The raised rim of the previous series is now developed into a hollow shell of considerable height, which tends to close over the drop. This shell or dome is a characteristic feature of all splashes made by large drops falling from a considerable height, and is extremely beautiful. In the splash at present under consideration it does not always succeed in closing permanently, but opens out as it subsides, and is followed by the emergence of the drop (Fig. 8). In Fig. 9 the return wave overwhelms the drop for an instant, but it is again seen at the summit of the column in Fig. 10.

SERIES IV.

The Splash of a Drop, followed in detail by Instantaneous Illumination.

Diameter of Drop, 1/4 inch. Height of Fall, 4 ft. 4 in.

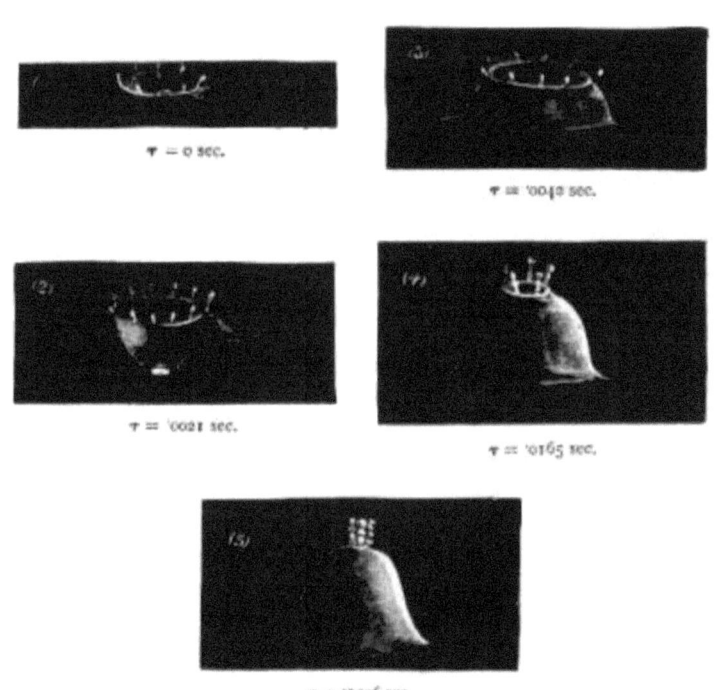

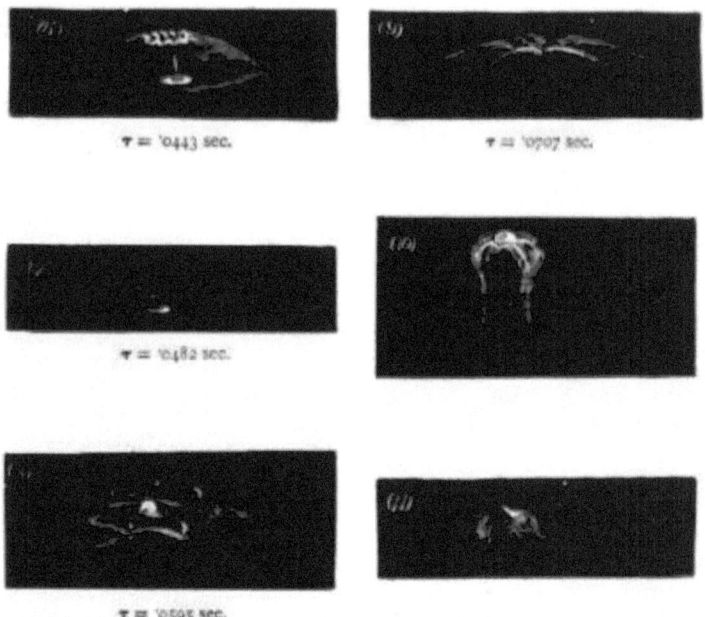

τ = ·0443 sec.

τ = ·0707 sec.

τ = ·0482 sec.

τ = ·0595 sec.

But on other occasions the shell or dome of Figs. 4 and 5 closes permanently over the imprisoned air, the liquid then flowing down the sides, which become thinner and thinner, till at length we are left with a large bubble floating on the water (see Series V.). It will be observed that the flow of liquid down the sides is chiefly along definite channels, which are probably determined by the arms thrown up at an earlier stage. The bubble is generally creased by the weight of the liquid along these channels. It must be remembered that the base of the bubble is in a state of oscillation, and that the whole is liable to burst at any moment, when such figures as 6 and 7 of the previous series will be seen.

SERIES V.

The Splash of a Drop, followed in detail by Instantaneous Illumination.

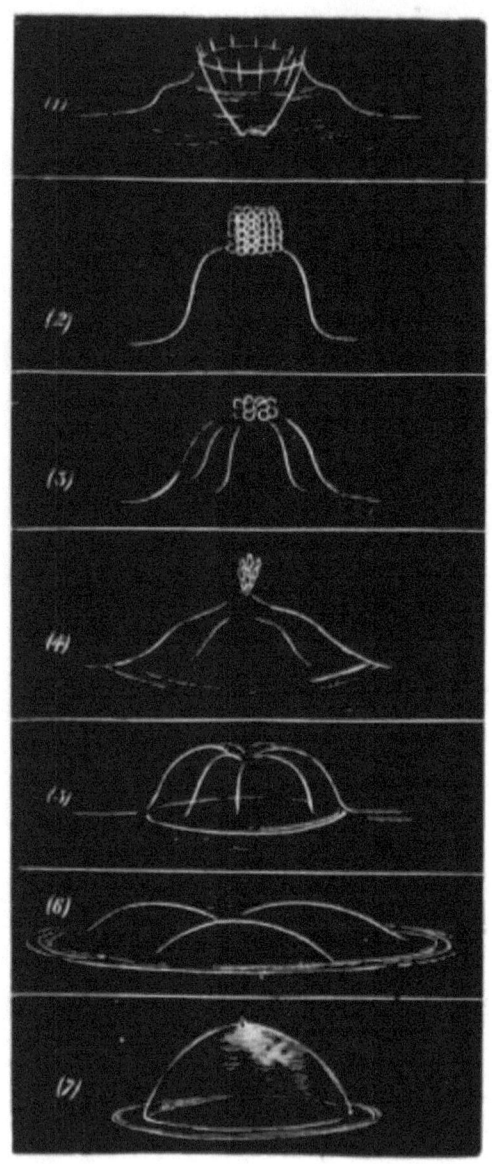

The Size of Drop and Height of Fall are the same as before, but the hollow shell (see figs. 4 and 5 of the previous Series) does not succeed in opening, but is left as a bubble on the surface. This explains the formation of bubbles when *big* rain-drops fall into a pool of water.]

Such is the history of the building of the bubbles which big rain-drops leave on the smooth water of a lake, or pond, or puddle. Only the bigger drops can do it, and reference to the number at the side of Fig. 5 of Series IV. shows that the dome is raised in about two-hundredths of a second. Should the domes fail to close, or should they open again, we have the emergent columns which any attentive observer will readily recognize, and which have never been better described than by Mr. R.L. Stevenson, who, in his delightful *Inland Voyage*, speaks of the surface of the Belgian canals along which he was canoeing, as thrown up by the rain into "an infinity of little crystal fountains."

Very beautiful forms of the same type indeed, but different in detail, are those produced by a drop of water falling into the lighter and more mobile liquid, petroleum.

It will now be interesting to turn to the splash that is produced when a solid sphere, such as a child's marble, falls into water.

I found to my great surprise that the character of the splash, at any rate up to a height of 4 or 5 feet, depends entirely on the state of the surface of the sphere. A polished sphere of marble about 0·6 of an inch in diameter, rubbed very dry with a cloth just beforehand and dropped from a height of 2 feet into water, gave the figures of Series VI., in which it is seen that the water spreads over the sphere so rapidly, that it is sheathed with the liquid even before it has passed below the general level of the surface. The splash is insignificantly small and of very short duration. [3] If the drying and polishing be not so perfect, the configurations of Series VII. are produced; while if the sphere be roughened with sandpaper, or *left wet*, Series VIII. is obtained, in which it will be perceived that, as was the case with the liquid drop, the water is driven away laterally, forming the ribbed basket-shaped hollow, which, however, is now prolonged to a great depth, the drop being followed by a cone of air, while the water seems to find great difficulty in wetting the surface completely. Part of this column of air was carried down at least 16 inches, and then only detached when the sphere struck the bottom of the vessel.

SERIES VI., VII.

Splash of a Solid Sphere (a marble 1/2 inch in diameter falling 2 feet into water).

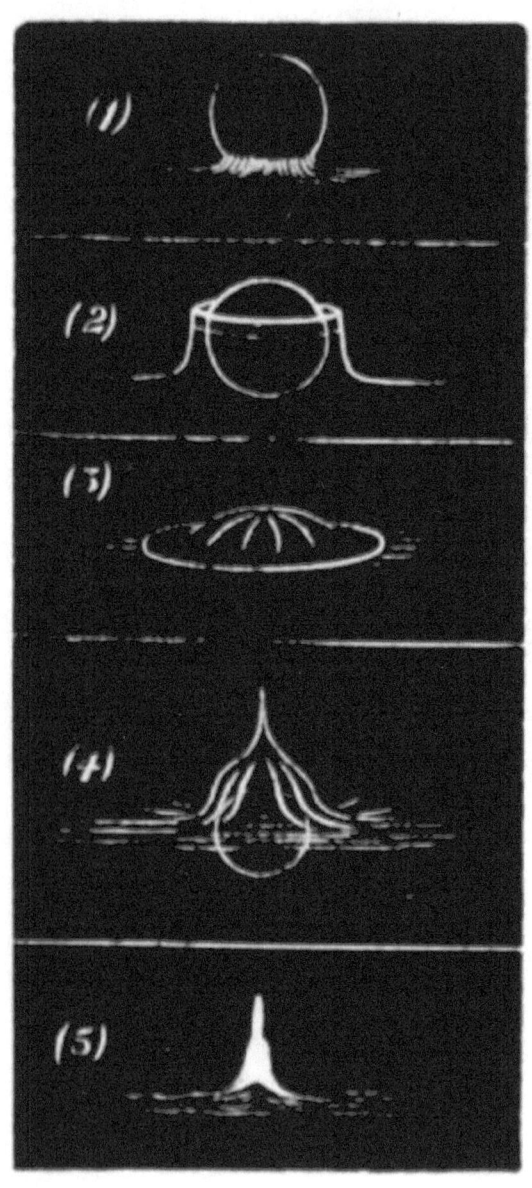

Series VI. When

the sphere is *dry* and *polished*.

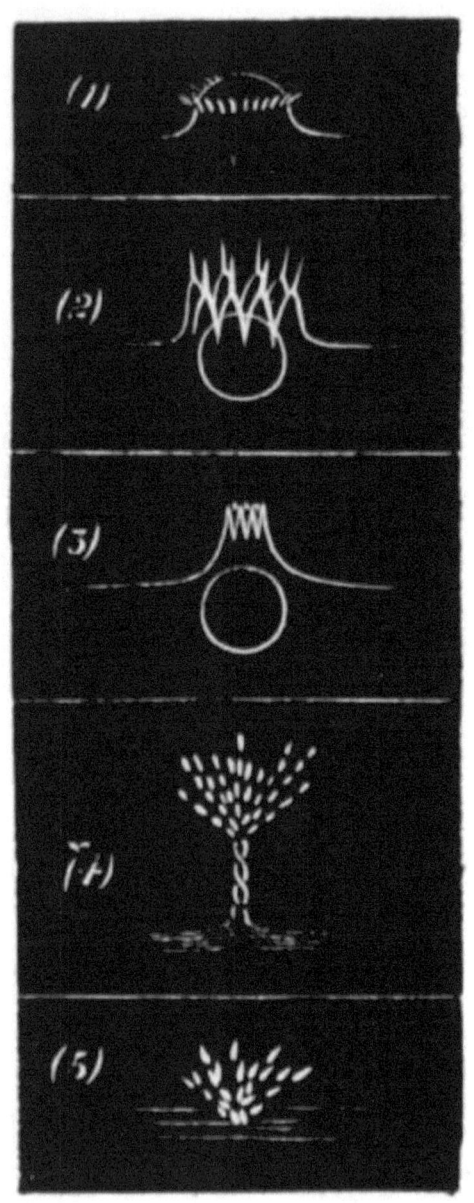

Series VII. When the

sphere is *not* well *dried* and *polished*.

SERIES VIII.

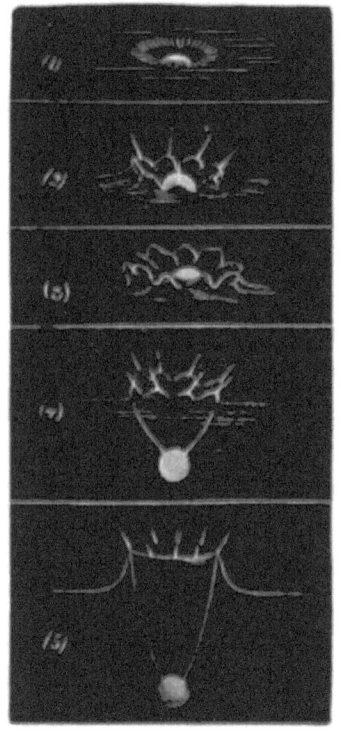
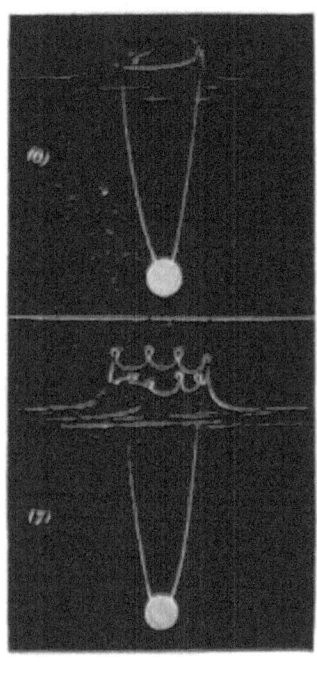

The Splash of a Solid Sphere — (continued.)
When the sphere is *rough* or *wet*.

SERIES IX.

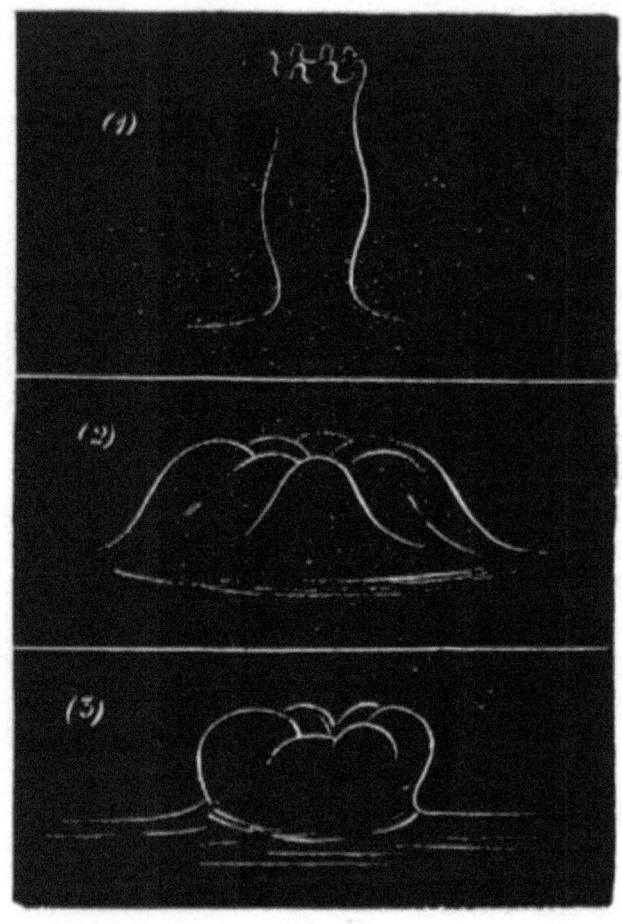

The Splash of a Solid Sphere—(continued.)
When the sphere is rough or wet, and falls above 5 feet.

Figs. 6 and 7 show the crater falling in, but this did not always happen, for the walls often closed over the hollow exactly as in Figs.

4 and 5 of Series IV. Meanwhile the long and nearly cylindrical portion below breaks up into bubbles which rise quickly to the surface.

By increasing the fall to 5 feet we obtain the figures of Series IX. The tube of Fig. 1 corresponds to the dome of Series IV. and V., and is not only elevated to a surprising height, but is also in the act of cleaving (the outline being approximately that of the unduloid of M. Plateau). Figs. 2 and 3 show the bubble formed by the closing up of this tube, weighed down in the centre as in Figs. 5 and 6 of Series V. Similar results were obtained with other liquids, such as petroleum and alcohol.

It is easy to show in a very striking manner the paramount influence of the condition of the solid surface. I have here a number of similar marbles; this set has been well polished by rubbing with wash leather. I drop them one by one through a space of about 1 foot into this deep, wide, cylindrical glass vessel, lighted up by a lamp placed behind it. You see each marble enters noiselessly and with hardly a visible trace of splash. Now I pick them out and drop them in again (or to save trouble, I drop in the place of these other wet ones), everything is changed. You see how the air is carried to the very bottom of the vessel, and you hear the "φλοῖσβος" of the bubbles as they rise to the surface and burst. These dry but rough marbles behave in much the same way.

Such are the main features of the Natural History of Splashes, as I made it out between thirteen and eighteen years ago. Before passing on to the photographs that I have since obtained, I desire to add a few words of comment. I have not till now alluded to any imperfections in the timing apparatus. But no apparatus of the kind can be absolutely perfect, and, as a matter of fact, when everything is adjusted so as to display a particular stage, it will happen that in a succession of observations there is a certain variation in what is seen. Thus the configuration viewed may be said to oscillate slightly about the mean for which the apparatus is adjusted. Now this is due both to small imperfections in the timing apparatus and to the fact that the splashes themselves do actually vary within certain limits. The reasons are not very far to seek. In the first place the rate of demagnetization of the electro-magnets varies slightly, being partly dependent on the varying resistance of the contacts of crossed wires, partly on the temperature of the magnet, which is affected by the length of time for which the current has been running. But a much more important reason is the variation of the slight adhesion of the drop to the smoked watch-glass that has supported it, and consequently of the oscillations to which, as we shall see, the drop is subjected as it descends. Thus the drop will sometimes strike the surface in a flattened form, at others in an elongated form, and there will be a difference, not only in the time of impact, but in the nature of the ensuing splash; consequently some judgment is required in selecting a consecutive series of drawings. The only way is to make a considerable number of drawings of each stage, and then to pick out a consecutive series. Now, whenever judgment has to be used, there is room for error of judgment, and moreover, it is impossible to put together the drawings so as to tell a consecutive story, without being guided by some theory, such as I have already sketched, as to the nature of the motion and the conditions that govern it. You will therefore be good enough to remember that this chronicle of the events of a tenth of a second is not a mechanical record but is presented by a fallible human historian, whose account, like that of any other contemporary observer, will be none the worse for independent confirmation. That confirmation is fortunately obtainable. In an attempt made eighteen years ago to photograph the splash of a drop of mercury, I was unable to procure plates sufficiently sensitive to respond to the very short exposures that were required, and consequently abandoned the endeavour. But in recent years plates of exquisite sensitiveness have been produced, and such photographs as those taken by Mr. Boys of a flying rifle bullet have shown that difficulties on the score of sensitiveness have been practically overcome. Within the last few weeks, with the valuable assistance of my colleague at Devonport, Mr. R.S. Cole, I have succeeded in obtaining photographs of various splashes. Following Prof. Boys' suggestion, we employed Thomas's cyclist plates, or occasionally the less sensitive "extra-rapid" plates of the same makers, and as a developer, Eikonogen solution of triple strength, in which the plates were kept for about 40 minutes, the development being conducted in complete darkness.

A few preliminary trials with the self-induction spark produced at the surface of mercury by the apparatus that you have seen at work, showed that the illumination, though ample for direct vision, was not sufficient for photography. When the current strength was increased, so as to make the illumination bright enough for the camera, then the spark became of too great duration, for it lasted for between 4 and 5 thousandths of a second, within which time there was very perceptible motion of the drop and consequent blurring. It was therefore necessary to modify the apparatus so as to employ a Leyden-jar spark whose duration was probably less than 10-millionths of a second. A very slight change in the apparatus rendered it suitable for the new conditions, but time does not permit me to describe the arrangements in detail. It is, however, less necessary to do so as the method is in all essentials the same as that described in this room two years ago by Lord Rayleigh in connection with the photography of a breaking soap-film. [4] I therefore pass at once to the photographs themselves.

The first two series (X. and XI.) may be described as shadow photographs; they were obtained by allowing a drop of mercury to fall on to the naked photographic plate itself, the illuminating spark being produced vertically above it, and they give only a horizontal section of the drop in various stages, revealing the form of the outline of the part in contact with the plate, but of course telling nothing about the shape of the parts above. The first series corresponds to a mercury splash very similar to that first described, and the second to the splash of a larger drop such as was not described. In each series, the tearing of the thin central film to which allusion was made is well illustrated. I think the first comment that any one would make is that the photographs, while they bear out the drawings in many details, show greater irregularity than the drawings would have led one to expect. On this point I shall presently have something to say.

SERIES X.

(1) *Instantaneous Shadow Photographs (life size) of the Splash of a Drop of Mercury falling 8 cm. on to the Photographic Plate.*

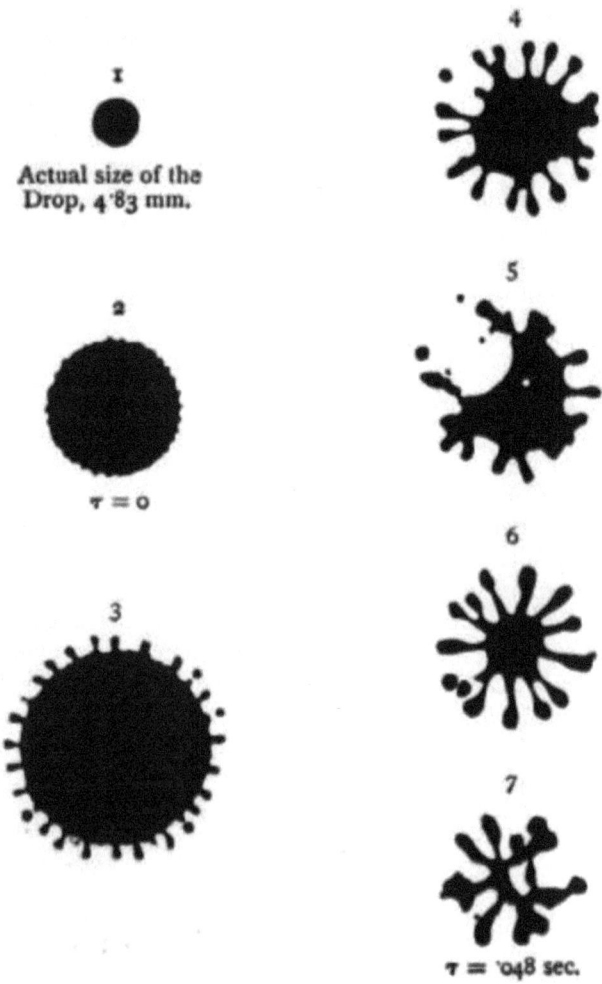

SERIES XI.

(2) *Instantaneous Shadow Photographs (life size) of the Splash of a Drop of Mercury falling 15 cm. on to Glass.*

1

Actual size, 4·83 mm.
in diameter.

2

$\tau = 0$ sec.

3

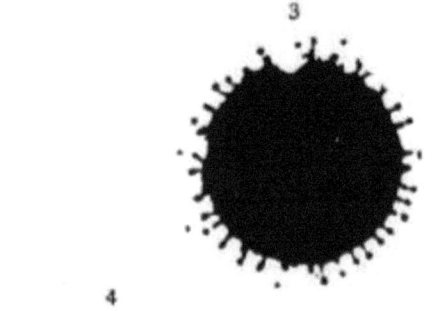

4   4A

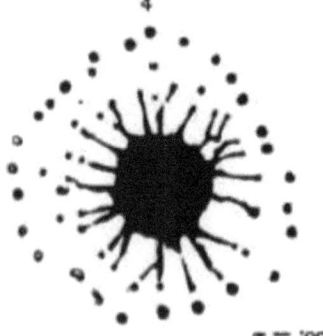 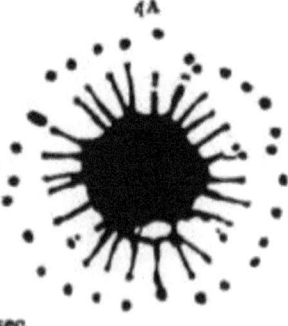

$\tau = $ ·0032 sec.

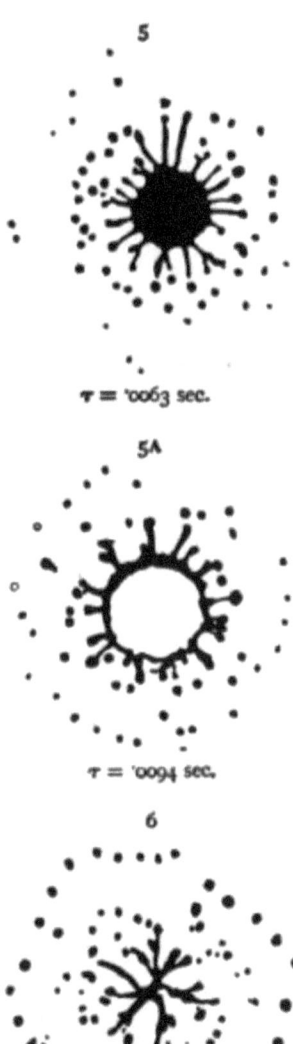

Comparing the first set of drawings (pp. 20-24) with the photographs of Series X., it will be seen that

	Photograph 2 corresponds to drawing 4 or 5
" 3	" " 9
" 4	" " 18
" 6	" " 20
" 7	" " 24

but the irregularity of the last photograph almost masks the resemblance.

SERIES XII.

Engravings from Instantaneous Photographs (16/17 of the real size) of the Splash of a Drop of Mercury, 4·83 mm. in diameter, falling 8·9 cm. on to a hard polished surface.

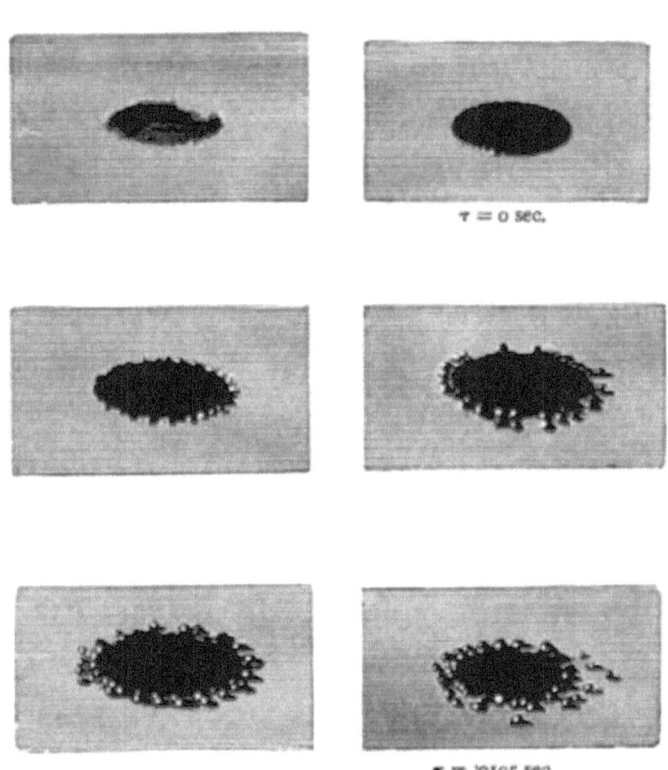

$\tau = 0$ sec.

$\tau = \cdot 0195$ sec.

Series XII. gives an objective view of a mercury splash as taken by the camera. Only the first of this series shows any detail in the interior. The polished surface of the mercury is, in fact, very troublesome to illuminate, and this splash proved the most difficult of all to photograph.

Series XIII. shows the splash of a drop of milk falling on to a smoked glass plate, on which it runs about without adhesion just as mercury would. Here there is more of detail. In Fig. 4 the central film is so thin in the middle that the black plate beneath it is seen through the liquid. In Fig. 8 this film has been torn.

Series XIV. exhibits the splash of a water drop falling into milk. The first four photographs show the oscillations of the drop about a mean spherical figure as it approaches the surface.

In the subsequent figures it will be noticed that the arms which are thrown up at first, afterwards segment into drops which fly off and subside (see Fig. 8), to be followed by a second series which again subside (Fig. 11), to be again succeeded by a third set. In fact, so long as there is any downward momentum the drop and the air behind it are penetrating the liquid, and so long must there be an upward flow of displaced liquid. Much of this flow is seen to be directed into the arms along the channels determined by the segmentation of the annular rim. This reproduction of the lobes and arms time after time on a varying scale goes far to explain the puzzling variations in their number which I mentioned in connection with the drawings. I had not, indeed, suspected this, which is one of the few new points that the photographs have so far revealed. [5]

SERIES XIII.

Engravings of Instantaneous Photographs (16/17 of the real size) of the Splash of a Drop of Milk falling 20 cm. on to smoked glass.

(It was not found possible to reproduce satisfactorily the missing figures of this series.)

1

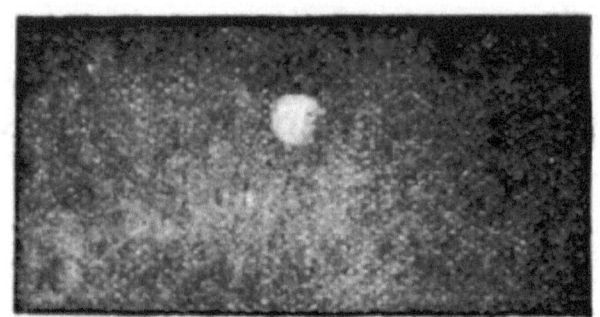

2

$\tau = 0$ sec.

3

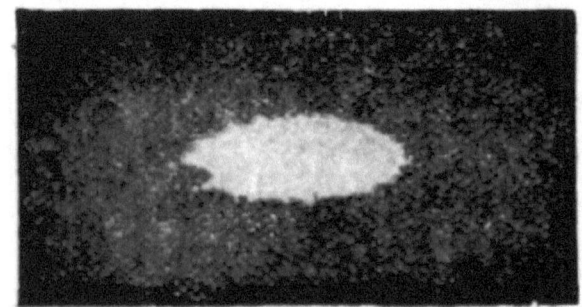

$\tau = \cdot 0025$ sec.

7

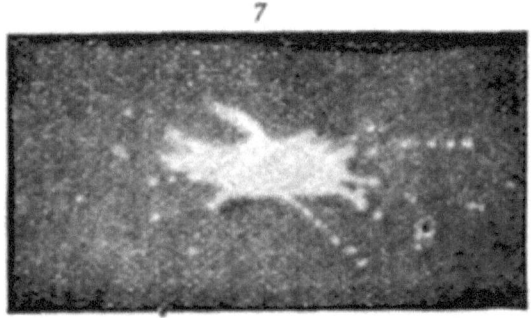

$\tau =$ ·0128 sec.

8

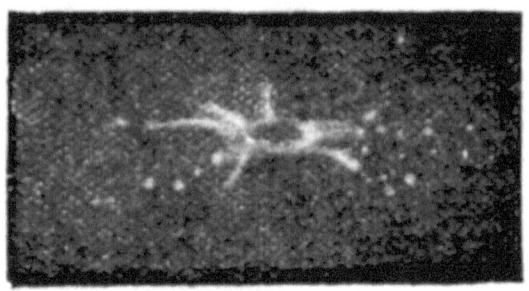

$\tau =$ ·0149 sec.

9

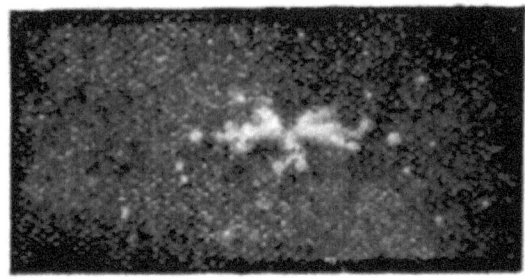

$\tau =$ ·0149 sec.

SERIES XIV.

Engravings of Instantaneous Photographs of the Splash of a Drop of Water falling 40 cm. into Milk.

Scale about 6/10 of actual size.

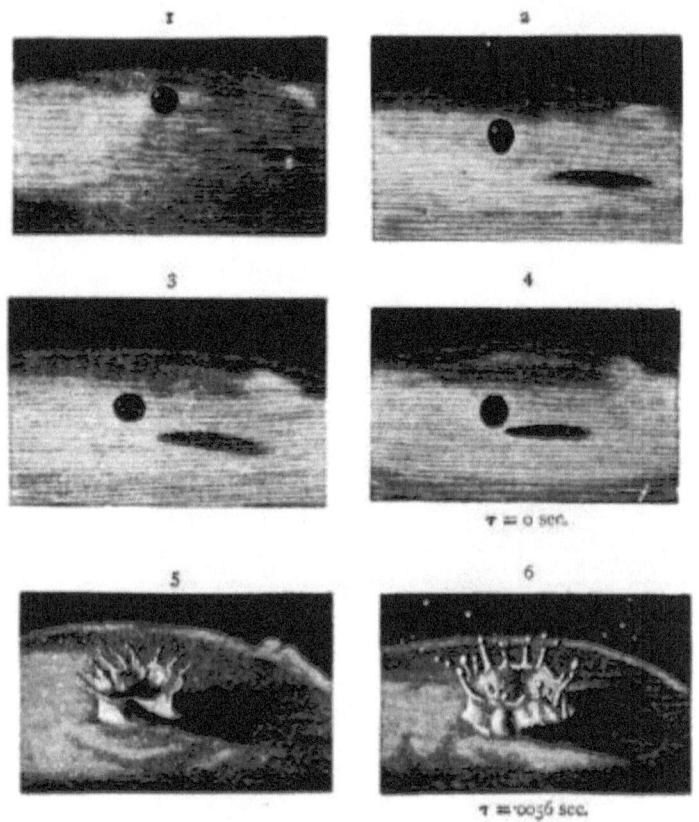

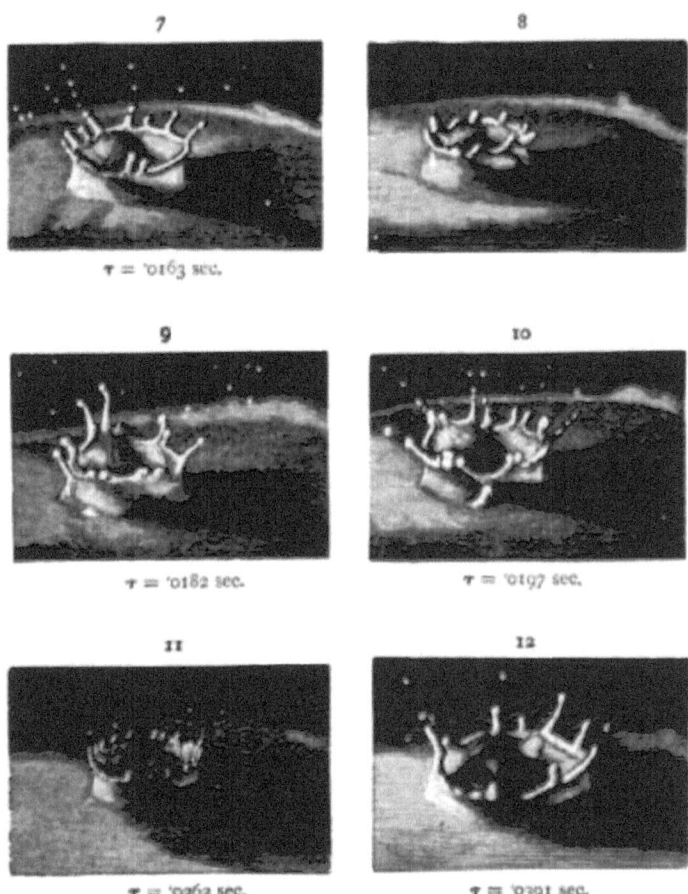

7
τ = ·0163 sec.

8

9
τ = ·0182 sec.

10
τ = ·0197 sec.

11
τ = ·0262 sec.

12
τ = ·0391 sec.

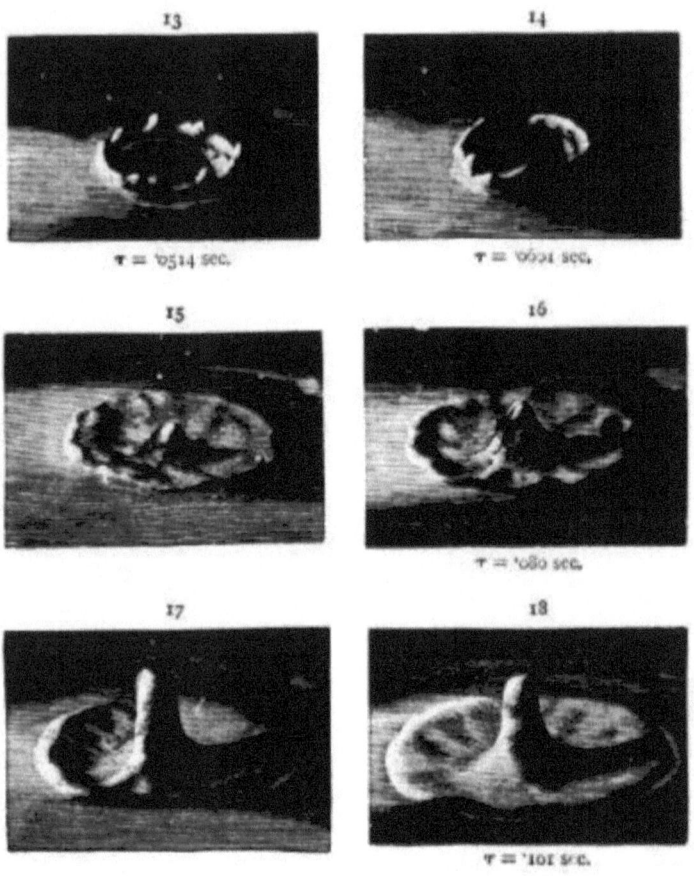

With respect to these photographs, [6] the credit of which I hope you will attribute firstly to the inventors of the sensitive plates, and secondly to the skill and experience of Mr. Cole, I desire to add that they are, as far as we know, the first really detailed objective views that have been obtained with anything approaching so short an exposure.

Even Mr. Boys' wonderful photographs of flying bullets were after all but shadow-photographs, and did not so strikingly illustrate the extreme sensitiveness of the plates, and I want you to distinguish between such and what (to borrow Mr. F.J. Smith's phrase) I call an "objective view."

It remains only to speak of the greater irregularity in the arms and rays as shown by the photographs. The point is a curious and interesting one. In the first place I have to confess that in looking over my original drawings I find records of many irregular or unsymmetrical figures, yet in compiling the history it has been inevitable that these should be rejected, if only because identical irregularities never recur. Thus the mind of the observer is filled with an ideal splash—an "Auto-Splash"—whose perfection may never be actually realized.

But in the second place, when the splash is nearly regular it is very difficult to detect irregularity. This is easily proved by projecting on the screen with instantaneous illumination such a photograph as that of Series X., Fig. 6. My experience is that most persons pronounce what they have seen to be a regular and symmetrical star-shaped figure, and they are surprised when they come to examine it by detail in continuous light to find how far this is from the truth. Especially is this the case if no irregularity is suspected beforehand. I believe that the observer, usually finding himself unable to attend to more than a portion of the rays in the system, is liable instinctively to pick out for attention a part of the circumference where they are regularly spaced, and to fill up the rest in imagination, and that where a ray may be really absent he prefers to consider that it has been imperfectly viewed.

This opinion is confirmed by the fact that in several cases, I have been able to observe with the naked eye a splash that was also simultaneously photographed, and have made the memorandum "quite regular," though the photograph subsequently showed irregularity. It must, however, be observed that the absolute darkness and other conditions necessary for photography are not very favourable for direct vision.

And now my tale is told, or rather as much of it as the limits of the time allowed me will permit. I think you will agree that the phenomena are very beautiful, and that the subject, commonplace and familiar though it is, has yet proved worthy of an hour's attention.

THE END.

Richard Clay & Sons, Limited, London & Bungay.

FOOTNOTES:

[1] See Worthington on the "Spontaneous Segmentation of a Liquid Annulus," *Proceedings Royal Society*, No. 200, p. 49 (1879).

[2] Readers who wish a more detailed account of a greater variety of splashes are referred to papers by the author. *Proceedings Royal Society*, vol. xxv. pp. 261 and 498 (1877); and vol. xxxiv. p. 217 (1882).

[3] Photographs obtained since this was written show that much may happen after the stages here traced.

[4] A detailed account of the optical, mechanical, and electrical arrangements employed, written by Mr. Cole, will be found in *Nature*, vol. i., p. 222 (July 5, 1894).

[5] The black streaks, seen especially in Figs. 11, 15, and 16, are due to particles of lamp-black carried down by the drop from the surface of the smoked watch-glass on which it rested.

[6] Three of these photographs, viz. Nos. 11, 12, and 17, are reproduced full size, as a frontispiece, by a *photographic* process, to enable the reader to form a more correct idea than can be gathered from the engravings, of the amount of detail actually obtained, though even in these reproductions much is inevitably lost.

THE ROMANCE OF SCIENCE.

Small post 8vo. Illustrated. Cloth boards.

	s.	d.
Coal; and What We get from It.		
By Professor Raphael Meldola, F.R.S., F.I.C.	2	6
Colour Measurement and Mixture.		
By Captain W. De W. Abney, C.B., R.E., F.R.S.	2	6
Diseases of Plants.		
By H. Marshall Ward, M.A., F.R.S., F.L.S.	2	6
Our Secret Friends and Foes.		

By Percy Faraday Frankland, Ph.D., B.Sc. (Lond.), F.R.S. Second Edition, revised and enlarged	3	0

Soap Bubbles and the Forces which Mould Them.

By C.V. Boys, A.R.S.M., F.R.S.	2	6

Spinning Tops.

By Professor J. Perry, M.E., D.Sc, F.R.S.	2	6

The Birth and Growth of Worlds.

By Professor Green, M.A., F.R.S.	2	6

The Making of Flowers.

By the Rev. Professor George Henslow, M.A., F.L.S., F.G.S.	2	6

The Story of a Tinder-Box.

By the late C. Meymott Tidy, M.B., M.S., F.C.S.	2	0

Time and Tide: a Romance of the Moon.

By Sir Robert S. Ball, F.R.S. Third Edition revised	2	6

PUBLICATIONS

OF THE

Society for Promoting Christian Knowledge.

PUBLICATIONS

OF

Society for Promoting Christian Knowledge.

Early Britain.

This is a Series of books which has for its aim the presentation of Early Britain at great Historic Periods.

Anglo-Saxon Britain.

By the late Grant Allen. With Map. Fcap. 8vo. *Cloth boards*, 2s. 6d.

Celtic Britain.

By Professor Rhys. With two Maps. Fcap. 8vo. *Cloth boards*, 3s.

Norman Britain.

By the Rev. W. Hunt. With Map. Fcap. 8vo. *Cloth boards*, 2s. 6d.

Post-Norman Britain.

By Henry G. Hewlett. Fcap. 8vo. *Cloth boards*, 3s.

Roman Britain.

By the Rev. Edward Conybeare. With Map. Fcap. 8vo. 3s. 6d.

Roman Roads in Britain.

By Thomas Codrington, M.Inst.C.E., F.G.S. With several Maps. Fcap. 8vo. *Cloth boards*, 5s.

THE HEATHEN WORLD AND ST. PAUL.

This Series is intended to throw light upon the Writings and Labours of the Apostle of the Gentiles.

Fcap. 8vo, cloth boards, 2s. each.

St. Paul in Damascus and Arabia.

By the late Rev. George Rawlinson, M.A., Canon of Canterbury. With Map.

St. Paul at Rome.

By the Very Rev. Charles Merivale, D.D., D.C.L., Dean of Ely. With Map.

St. Paul in Asia Minor and at the Syrian Antioch.

By the late Rev. E.H. Plumptre, D.D. With Map.

ANCIENT HISTORY FROM THE MONUMENTS

This Series of Books is chiefly intended to illustrate the Sacred Scriptures by the results of recent Monumental Researches in the East.

Fcap. 8vo, cloth boards, 2s. each.

Assyria, from the Earliest Times to the Fall of Nineveh.

By the late George Smith, of the British Museum. New and Revised Edition by the Rev. Professor A.H. Sayce.

Sinai, from the Fourth Egyptian Dynasty to the Present Day.

By the late Henry S. Palmer, Major R.E., F.R.A.S. With Map. A New Edition, revised throughout. By the Rev. Professor A.H. Sayce.

Babylonia (The History of).

By the late George Smith. Edited by the Rev. Professor A.H. Sayce.

Persia, from the Earliest Period to the Arab Conquest.

By the late W.S.W. Vaux, M.A., F.R.S. A New Edition. Revised by the Rev. Professor A.H. Sayce.

BOOKS BY THE AUTHOR OF

"The Chronicles of the Schönberg-Cotta Family."

"By the Mystery of Thy Holy Incarnation."

Post 8vo. *Cloth boards*, 1s.

"By Thy Cross and Passion."

Thoughts on the words spoken around and on the Cross. Post 8vo. *Cloth boards*, 1s.

"By Thy Glorious Resurrection and Ascension."

Easter Thoughts. Post 8vo. *Cloth boards*, 1s.

"By the Coming of the Holy Ghost."

Thoughts for Whitsuntide. Post 8vo. *Cloth boards*, 1s.

The True Vine.

Post 8vo. *Cloth boards*, 1s.

The Great Prayer of Christendom.

Post 8vo. *Cloth boards*, 1s.

An Old Story of Bethlehem.

One link in the great Pedigree. Fcap. 4to, with six plates beautifully printed in colours. *Cloth boards*, 2s.

Three Martyrs of the Nineteenth Century.

Studies from the Lives of Livingstone, Gordon, and Patteson. Crown 8vo. *Cloth boards*, 2s. 6d.

Martyrs and Saints of the First Twelve Centuries.

Studies from the Lives of the Black-letter Saints of the English Calendar. Crown 8vo. *Cloth boards*, 3s. 6d.

Against the Stream.

The Story of an Heroic Age in England. With eight page woodcuts. Crown 8vo. *Cloth boards*, 2s. 6d.

Conquering and to Conquer.

A Story of Rome in the days of St. Jerome. With four page woodcuts. Crown 8vo. *Cloth boards*, 2s.

Lapsed not Lost.

A Story of Roman Carthage. Crown 8vo. *Cloth boards*, 2s.

Sketches of the Women of Christendom.

Crown 8vo. *Cloth boards*, 2s. 6d.

Thoughts and Characters.

Being Selections from the Writings of the Author of "The Schönberg-Cotta Family." Crown 8vo. *Cloth bds.*, 2s. 6d.

The Fathers for English Readers.

Fcap. 8vo, cloth boards, 2s. each.

Boniface.

By the Rev. Canon Gregory Smith, M.A. (1s. 6d.)

Clement of Alexandria.

By the Rev. F.R. Montgomery Hitchcock. (3s.)

Gregory the Great.

By the late Rev. J. Barmby, B.D.

Leo the Great.

By the Right Rev. C. Gore, D.D.

Saint Ambrose: his Life, Times, and Teaching.

By the Ven. Archdeacon Thornton, D.D.

Saint Athanasius: his Life and Times.

By the Rev. R. Wheler Bush. (2s. 6d.)

Saint Augustine.

By the late Rev. E.L. Cutts, D.D.

Saint Basil the Great.

By the Rev. Richard T. Smith, B.D.

Saint Bernard: Abbot of Clairvaux, A.D. 1091-1153.

By the Rev. S.J. Eales, M.A., D.C.L. (2s. 6d.)

Saint Hilary of Poitiers, and Saint Martin of Tours.

By the Rev. J. Gibson Cazenove, D.D.

Saint Jerome.

By the late Rev. Edward L. Cutts, D.D.

Saint John of Damascus.

By the Rev. J.H. Lupton, M.A.

Saint Patrick: his Life and Teaching.

By the Rev. E.J. Newell, M.A. (2s. 6d.)

Synesius of Cyrene, Philosopher and Bishop.

By Alice Gardner.

The Apostolic Fathers.

By the Rev. Canon H.S. Holland.

The Defenders of the Faith; or, The Christian Apologists of the Second and Third Centuries.

By the Rev. F. Watson, D.D.

The Venerable Bede.

By the Right Rev. G.F. Browne, D.D.

Diocesan Histories.

Bath and Wells.

By the Rev. W. Hunt. With Map, 2s. 6d.

Canterbury.

By the late Rev. R.C. Jenkins. With Map, 2s. 6d.

Carlisle.

By the late Richard S. Ferguson. With Map, 2s. 6d.

Chester.

By the Rev. Rupert H. Morris, D.D. With Map, 3s.

Chichester.

By the late Very Rev. W.R.W. Stephens. With Map and Plan, 2s. 6d.

Durham.

By the Rev. J.L. Low. With Map and Plan, 2s. 6d.

Hereford.

By the Rev. Canon Phillott. With Map, 3s.

Lichfield.

By the late Rev. W. Beresford. With Map, 2s. 6d.

Lincoln.

By the late Canon E. Venables, and the late Ven. Archdeacon Perry. With Map, 4s.

Llandaff.

By the Rev. E.J. Newell, M.A. With Map, 3s. 6d.

Norwich.

By the Rev. A. Jessop, D.D. With Map, 2s. 6d.

Oxford.

By the Rev. E. Marshall. With Map, 2s. 6d.

Peterborough.

By the Rev. G.A. Poolem, M.A. With Map, 2s. 6d.

Rochester.

By the Rev. A.J. Pearman, M.A. With Map, 4s.

Salisbury.

By the Rev. W.H. Jones. With Map and Plan, 2s. 6d.

Sodor and Man.

By A.W. Moore, M.A. With Map, 3s.

St. Asaph.

By the Ven. Archdeacon Thomas. With Map, 2s.

St. Davids.

By the Rev. Canon Bevan. With Map, 2s. 6d.

Winchester.

By the Rev. W. Benham, B.D. With Map, 3s.

Worcester.

By the Rev. J. Gregory Smith, M.A., and the Rev. Phipps Onslow, M.A. With Map, 3s. 6d.

York.

By Rev. Canon Ornsby, M.A., F.S.A. With Map, 3s. 6d.

Miscellaneous Publications.

A Dictionary of the Church of England.

By the late Rev. Edward L. Cuttsm, D.D. With numerous Woodcuts. Crown 8vo. 5s. Third Edition, revised.

Aids to Prayer.

By the Rev. Daniel Moore. Printed in red and black. Post 8vo. 1s. 6d.

Being of God (Six Addresses on the).

By C.J. Ellicott, D.D., Bishop of Gloucester. Post 8vo. 1s. 6d.

Bible Places; or, The Topography of the Holy Land.

By the Rev. Canon Tristram. With Map and numerous Woodcuts. New Edition. Crown 8vo. 5s.

Called to be Saints.

The Minor Festivals Devotionally Studied. By the late Christina G. Rossetti, Author of "Seek and Find." Post 8vo. 3s. 6d.

Case for "Establishment" stated (The).

By the Rev. T. Moore, M.A. Post 8vo. *Paper boards*, 6d.

Christians under the Crescent in Asia.

By the late Rev. E.L. Cutts, D.D. With numerous Illustrations. Crown 8vo. 5d.

Daily Readings for a Year.

By Elizabeth Spooner. Crown 8vo. 3s. 6d.

Devotional (A) Life of Our Lord.

By the late Rev. Edward L. Cutts, D.D., Author of "Pastoral Counsels," &c. Post 8vo. 5s.

Golden Year, The.

Thoughts for every month. Original and Selected. By Emily C. Orr, Author of "Thoughts for Working Days." Printed in red and black. Post 8vo. 1s.

Gospels (The Four).

Arranged in the Form of an English Harmony, from the Text of the Authorized Version. By the late Rev. J.M. Fuller. With Analytical Table of Contents and Four Maps. 1s.

Holy Eucharist, The Evidential Value of the.

Being the Boyle Lectures for 1879 and 1880. By the late Rev. G.F. Maclear, D.D. Crown 8vo. *Cloth boards*, 4s.

Land of Israel (The).

A Journal of Travel in Palestine, undertaken with special reference to its Physical Character. By the Rev. Canon Tristram. With Two Maps and numerous Illustrations. Large Post 8vo. *Cloth boards*, 10s. 6d.

Lectures on the Historical and Dogmatical Position of the Church of England.

By the Rev. W. Baker, D.D. Post 8vo. *Cloth boards*, 1s. 6d.

Paley's Evidences.

A New Edition, with Notes, Appendix, and Preface. By the Rev. E.A. Litton. Post 8vo. *Cloth boards*, 4s.

Paley's Horæ Paulinæ.

A New Edition, with Notes, Appendix, and Preface. By the late Rev. J.S. Howson, D.D. Post 8vo. *Cloth boards*, 3s.

Peace with God.

A Manual for the Sick. By the Rev. E. BurbidGE, M.A. Post 8vo. *Cloth boards*, 1s. 6d.

Plain Words for Christ.

Being a Series of Readings for Working Men. By the late Rev. R.G. Dutton. Post 8vo. *Cloth boards*, 1s.

Readings on the First Lessons for Sundays and Chief Holy Days.

According to the New Table. By the late Rev. Peter Young. Crown 8vo. *In two volumes*, 5s.

Reflected Lights.

From Christina Rossetti's "The Face of the Deep." Selected and arranged by W.M.L. Jay. Small post 8vo. *Cloth boards*, 2s. 6d.

Some Chief Truths of Religion.

By the late Rev. Edward L. Cutts, D.D., Author of "St. Cedd's Cross," &c. Crown 8vo. *Cloth boards*, 2s. 6d.

Thoughts for Men and Women.

The Lord's Prayer. By Emily C. Orr. Post 8vo. *Limp cloth*, 1s.

Thoughts for Working Days.

Original and Selected. By Emily C. Orr. Post 8vo. *Limp cloth*, 1s.

Time Flies; a Reading Diary.

By the late Christina G. Rossetti. Post 8vo. *Cloth boards*, 3s. 6d.

Turning Points of English Church History.

By the late Rev. Edward L. Cutts, D.D. Crown 8vo. *Cloth boards*, 3s. 6d.

Turning Points of General Church History.

By the late Rev. E.L. Cutts, D.D., Author of "Pastoral Counsels." Crown 8vo. *Cloth boards*, 4s.

Verses.

By the late Christina G. Rossetti. Printed on hand-made paper. Small Post 8vo. *Cloth*, 3s. 6d.

LONDON:
NORTHUMBERLAND AVENUE, CHARING CROSS, W.C.;
48, QUEEN VICTORIA STREET, E.C.
BRIGHTON: 129, North Street.

www.ingramcontent.com/pod-product-compliance
Lightning Source LLC
Chambersburg PA
CBHW030456220526
45464CB00006B/2560